Design and Make

Polyester Jewelry

Jan and Ted Arundell

J M Dent & Sons Ltd   London

Erratum: On page 22, figures 16 and 17 have been transposed; the illustrations (but not the illustration numbers) should therefore be switched round.

The authors would like to thank Sidney Cleaver of Trylon Ltd for his valuable technical assistance, Annie Whitaker for her patience and help in preparing and typing the manuscript, and Brigette who 'appears' in many of the photographs.

First published 1975

Made in Great Britain
for J. M. DENT & SONS LTD
Aldine House · Albemarle Street · London

Filmset by BAS Printers Limited, Wallop Hampshire

ISBN 0 460 04121 5

Contents

1

Introduction

If you have been enchanted by the way the artist or craftsman can create beautiful objects with a few simple strokes of brush or knife, you will be delighted by the results obtained from using polyester resin. By adding a few drops of catalyst to a thin syrup-like substance, a material can be made that possesses all the brilliance of glass.

Such 'magic' may be employed to produce jewelry of the kind illustrated in this book and it can be put to work in ways that remain for you to discover. After doing some of the projects, perform a little sorcery of your own and surprise yourself with your originality.

American terms are given in square brackets where they differ from the British.

2

3

4

6   5

Using the Book

6

First read through the yellow pages to get a general picture of what is involved in designing and making polyester jewelry. When you have done this you will be ready to start the projects. Wherever necessary, reference is given to the information contained in the yellow pages. If you are uncertain of how to perform an operation, you simply turn to the page indicated by the reference number and read the relevant instructions. The more skilled you become, the less you will need to refer to the yellow pages. Eventually you will be able to design and make original jewelry with only occasional reference to the book.

The easier projects are towards the front of the book and beginners are advised to start on these. If you feel confident, you may choose to do the more advanced projects, in which case you will find that each project is self-contained and provides enough information to carry it through.

Polyester Resin

Polyester resin is most commonly used as a binding agent in fibre-glass construction, in which form it is employed to make a large range of products such as motor car bodies, boat hulls, swimming pool linings etc. There are other less-known applications of polyester resin, such as its use as an embedding medium for scientific specimens. This use is very much akin to the way in which we shall be employing it.

Basic Materials and Equipment

Below is a list of materials and equipment which you will need to start making polyester jewelry.

Resin — we recommend that you use the pre-activated (PA) variety, p. 17
Barrier [protective] cream, p. 10
Old newspapers, p. 10
Liquid cleaner (based on acetone), p. 10
Tissues, p. 11
Small spirit level [called simply a level in USA], p. 12
Assorted polythene [polyethylene] bottles, p. 13
Assorted polythene lids, p. 13
Containers, p. 14
Opaque polyester pigments, p. 16

Translucent polyester pigments, p. 16
Catalyst, p. 17
Calibrated catalyst dispenser and/or catalyst dropper, p. 18, fig. 13
Calibrated measuring cylinder and/or paper cups, p. 18, fig. 13
Stirring implements, p. 19
Silicon carbide paper, p. 21
Jeweller's findings, e.g. ring mountings (or shanks), cuff-link fittings, brooch pins, bolt rings [spring rings], jump rings etc.

None of these items are expensive and some can be retrieved from domestic waste. For the names and addresses of suppliers refer to p. 64.

Preparation for Work

At all times maintain a high standard of 'housekeeping'. Try to keep spillage to a minimum and always clean up waste material immediately.

1 Care of yourself

a Apply the barrier [protective] cream to the hands (and arms if they are likely to come into contact with the chemicals).

b Do not allow chemicals to come into unnecessary contact with the skin.

c Take the utmost care with the catalyst (Hydrogen Peroxide). If it is splashed on the skin, wash off immediately with a lot of water. If eyes should be splashed, wash with 2% solution of sodium bicarbonate and get immediate medical attention. Always keep the catalyst in the recommended type of dispenser from which it should be poured with care.

d When work is complete, rub the hands with special cleansing cream and then wash them with soap and water.

e A very small percentage of people are allergic to polyester resin. If any kind of skin irritation should develop after taking all the above precautions, the sufferer should completely abandon work with these materials. (However, people who are allergic to polyester resin may wear jewelry made from it provided all areas of the jewelry likely to come into contact with the skin for any length of time are given a coat of clear polyurethane or nail varnish.)

f Always supervise young children when they are using polyester resin.

g Always work in a ventilated room, i.e. with at least one window open.

h Do not wear rings, watches etc., whilst working with the chemicals.

2 Care of clothes

a Always cover clothes with an overall or wear garments kept specially for this work.

b Should resin be accidentally spilled on clothes, it may be possible to clean it off with liquid cleaner (based on acetone) applied with a paper tissue. Make a small test on an unimportant part of the garment, as the cleaner completely dissolves certain man-made fibres such as Tricel.

3 Care of tables

a Completely cover all working surfaces with several layers of newspaper, which should be thrown away at the end of each job.

b It is not advisable to use a plastic table-covering, such as a sheet of polythene, as it would be difficult to keep this really clean.

4 Care of floors
Work in a room with wooden, concrete, pottery tile or earth floor. Never work over plastic tiles as they are very badly affected by some of the chemicals used.

5 Care of utensils and tools
a All tools and non-disposable containers etc, must be kept clean if good results are to be obtained. Clean everything with tissues soaked in liquid cleaner. Throw away the tissues after use.

b Never pour surplus materials down the sink as they will block the drain.

c Collect surplus material, disused containers, paper tissues and table coverings for disposal in a receptacle which can be closed.

d Dispose of waste material regularly.

6 Safety
a Most of the chemicals referred to in this book burn very readily. Do not allow any of them to come near to a naked flame or become overheated.

b Care should be taken when disposing of surplus resin which is still curing. If the resin is more than about 1 cm [$\frac{1}{2}$ in] thick, its temperature will rise to about 150°C [302°F] during curing. This material should not, under any circumstances, be put into a waste container until it has cooled off.

c Do not store materials in extreme temperatures. 18°C [approx. 65°F] is the recommended storage temperature.

d If the catalyst should freeze it will crystallize and must be allowed to thaw before it is handled. Should the liquid have evaporated and crystals remain, dispose of them carefully, avoiding direct contact with the hands.

e The catalyst should be stored in a separate fire-proof compartment out of direct sunlight. Always ensure that the container is not more than $\frac{2}{3}$ full.

f To reduce the chance of the wrong chemical being used, make sure that all containers are clearly and boldly labelled. It is helpful to use containers of different shapes and/or colours. The catalyst must be kept in its original container or a recom- mended type of dispenser. NB Many plastics will react with the chemicals, therefore avoid using plastic containers other than those made of polythene and p.v.c.

g Keep a fire blanket or fire extinguisher to hand.

h *Never, never* smoke whilst working with these materials.

Checking and Levelling a Surface

7 When casting in a mould one relies on gravity to ensure that the back and front faces of the cast are parallel. Therefore, until the resin has gelled, it is important that the mould rests on a perfectly level surface.

If, after testing with a spirit level [level], you do not have an area that is suitable, place a flat, rigid board on a convenient surface and prop it to a level position with small items of suitable thickness. Washers are useful for this purpose as you can adjust the level by varying their number and position (fig. 8). It is good practice to place the specially levelled surface away from the general work area to avoid accidentally knocking the filled moulds.

8

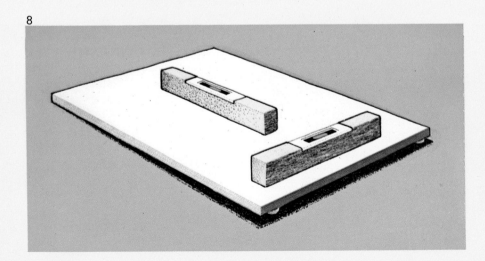

Moulds

9

8 Most of the pieces of jewelry in this book are moulded in polythene lids and suitably cut up polythene bottles of detergents, shampoos, etc. retrieved from domestic waste. A few pieces have been moulded in items

10

bought specially for the purpose, i.e., measuring spoons, ice cube containers, etc. Polythene is a material to which cured polyester will not stick. This quality, combined with the flexibility of polythene, makes it easy to release items cast in moulds made from it. Polythene, unlike many other plastics, will not react with polyester resin. [Note for US readers: polythene is known in the USA as polyethylene.]

Retrieved lids and bottles etc. that can be used as moulds provide a wealth of shapes and sizes for producing jewelry. It is best to make a collection of 'moulds' and design your jewelry to fit them. This approach produces more interesting results, more quickly, than trying to find a mould to suit a design which has already been worked out. You do not need moulds identical to those in the projects. Differences will give a personal touch to your jewelry.

9 Cutting up polythene bottles for use as moulds
i Cut away the unwanted part of the bottle leaving a mould about $2\frac{1}{2}$ cm [1 in] deep (fig. 9a).
ii Make four 1 cm [or roughly $\frac{1}{2}$ in] vertical cuts, equal distances apart, as shown in fig. 9b.
iii Release two oblongs of material by making two horizontal cuts around the side of the mould (fig. 9c).

Handling Polyester Resin

11

10 Cleaning polyester resin from containers is an unpleasant job, so whenever possible use disposable containers for mixing. If non-disposable containers have to be used, they must be very carefully cleaned afterwards to avoid spoiling the next mix.

11 Special containers
Wax paper containers are sold especially for use with polyester resin and can be obtained in several sizes. It is useful to have at least two sizes, i.e. 20 oz and 8 oz calibrated cups. As these cost money it is best to use them for measuring and handling resin and to rely on retrieved domestic containers for holding material that is to be catalysed.

12 Retrieved containers
Living in a consumer society we are particularly fortunate in having access to an almost limitless array of items that are well suited to our present exercise. Glass jars, polythene bottles and tin cans, all of which are usually thrown away, can be given a new lease of life. The smaller containers are generally the most useful because, storage apart, the quantities of resin to be used at any one time are small. The best policy is to start a collection well before beginning serious jewelry making.

a *Tin cans* The cans must be opened neatly. The simple lever action tin openers make a jagged edge, rendering the can unsuitable for our purpose. Provided the cans are rust free and properly cleaned they are ideal for containing polyester resin.

b *Plastic containers* Plastic egg cartons are useful for holding small amounts of resin to be catalysed. The individual cases can be separated with a pair of scissors. They may fall on their sides when empty but will stand upright when filled with resin. However, egg cases should only be employed to hold resin for immediate use, as resin left standing in the cases will dissolve them. It is worth bearing with this disadvantage as it is difficult to come by other retrieved containers of this size.

Many plastic containers dissolve when left in contact with polyester resin, therefore only use

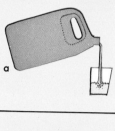

12 c

those made from polythene or p.v.c. for holding resin for periods longer than a few minutes.

c *Glass jars* These are the most re-usable of the containers, as they are usually the easiest to clean. It is also possible to see clearly material contained in them, which is an advantage when mixing in catalyst and assessing the hue of added colouring.

13 How to make tins and paper cups pour satisfactorily

Where material is being poured from a larger to a smaller vessel, a simple way to increase control of the pouring is to reduce the flow area of the material leaving the containing vessel. All you need to do is squeeze the rim of the vessel so that it has a lip

like a jug (fig. 11). The procedure is then to pour into the back of the small 'jug' from the pinched-in lip of the large 'jug' (fig. 12c). A vessel shaped in this way is particularly useful when pouring resin into the egg cases referred to on p. 14.

14 Pouring

Resin can be bought in various quantities, the larger the amount the cheaper the price per lb. Should you wish to take advantage of quantity buying you would be advised to pour some resin from the large container in which it is delivered, into a smaller container for your immediate use.

When transferring resin from one vessel to another, care should be taken to include as little air as possible in the material (see p. 19 ‡ 25). The best way of doing this is:

i Stand the vessel that is to receive the resin and the vessel holding the resin side by side.

ii Tip both vessels so that their rims meet at the angle that permits the resin to flow gently down the inner side of the receiving vessel (fig. 12b and c).

iii Continue pouring until the receiving vessel is $\frac{2}{3}$ full (do not fill above this level or it will be difficult to control the flow of resin when it is poured out again).

Using Coloured Pigments

jars or plastic egg cartons. This will ensure that light enters the mix through the sides of the containers in the same way as it will enter the finished item. If the containers are opaque the coloured resin will read very much darker than if light is allowed to enter the mix freely and you will have difficulty in judging the true hue of the material.

The surest way of getting the desired hue is to mix the pigment into the resin in small amounts until you have the right intensity of colour.

15 Translucent polyester pigments

The beautiful optical qualities of cast polyester can be enhanced by using special polyester pigments, which can be bought inexpensively in small amounts for making jewelry.

16 Mixing translucent polyester pigments

With as few as four translucent colours you can achieve excellent results. Like paints these may be mixed together to create additional colours. The hue may be varied in the same way as with water-colours, with the difference that the pigments are mixed in resin and not water. In fact, the effects that can be achieved are remarkably similar to those produced by water-colour, although the brilliance created by polyester pigments is more intense.

In order to obtain a 'true reading' of the hue of the finished piece of jewelry, mix the pigment and resin together in transparent or translucent containers, e.g. glass

17 Opaque polyester pigments

To reflect light and/or colour through the translucent material, most of the items illustrated are either partly or fully backed with opaque polyester.

Opacity is achieved by adding opaque pigment to the resin. Like translucent pigments these may be obtained in small quantities. You would be advised to buy at least five colours — black, white and the three primaries, red, blue and yellow. By mixing these together your 'palette' can be extensive.

18 Mixing opaque polyester pigments

These may be mixed together with resin in the same way as translucent pigments except that there is no need to use transparent containers to hold the material. In the same way that translucent pigments resemble water-colours, so opaque pigments resemble gouache or poster colours. If you have used either of the latter you will feel at home mixing opaque polyester pigments.

NB The amount of pigment used should not exceed 10% of the total mix.

Processing Polyester Resin

19 For our purpose polyester exists in four conditions: liquid, gelled, set and cured.

i Polyester is supplied as a *liquid.*
ii Polyester *gells* (becomes jelly-like) in 15 to 30 minutes after being catalysed.
iii Polyester is *set* when it becomes an inflexible solid – about 2 hours after being catalysed.
iv Polyester is *cured* when it has thoroughly hardened – between 24 hours and 7 days after being catalysed.
(The above times are approximate. They are based on using the amounts of catalyst given in fig. 13 at room temperature.)

In time, polyester resin will set hard without the use of a catalyst. For this reason do not buy more resin than you can use in a period of six months. It is not practical to wait for the resin to set of its own accord, therefore a catalyst must be used to speed up the process. For making jewelry we recommend the liquid type of catalyst (Organic Peroxides). It is best to buy the catalyst in small amounts and to use it within a reasonable time, as the longer it is kept the less effective it becomes.

Always use a pre-activated (PA) polyester resin, otherwise you will have to use an activator as well as a catalyst. This should be avoided because it makes the mixing of the materials un-necessarily complicated. It can also be dangerous, as the catalyst and activator can react violently if mixed together.

NB There are many types of resin. We have found a general purpose laminating PA resin perfectly satisfactory.

Measuring
Resin and Catalyst

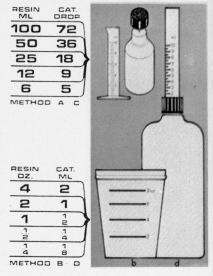

RESIN ML	CAT. DROP
100	72
50	36
25	18
12	9
6	5

METHOD A C

RESIN OZ.	CAT. ML
4	2
2	1
1	$\frac{1}{2}$
$\frac{1}{2}$	$\frac{1}{4}$
$\frac{1}{4}$	$\frac{1}{8}$

METHOD B D

13

20 *Resin* may be measured in two ways:
In a polythene cylindrical container calibrated up to 10 millilitres [10 cc] (fig. 13a),
or
in a paper cup calibrated from 2 oz to 8 oz (fig. 13b).

21 *Catalyst* may be measured in two ways:
Drop by drop from a small dropper bottle (fig. 13c),
or
in a polythene bottle fitted with a measuring cylinder calibrated up to 10 ml [10 cc] (fig. 13d).
NB The cylinder is filled by squeezing the bottle.

When measuring the small amounts generally needed to make jewelry, we recommend using A/C (fig. 13). Fig 13 gives the approximate amounts of liquid catalyst to be used with various quantities of pre-activated resin. The age of the materials and the temperature of the room in which they are stored and mixed, will affect the time it takes the resin to gel. (The warmer the room the quicker the resin will gel. However, do not catalyse resin at a room temperature lower than 15°C. [60°F], otherwise the item cast will be undercured.) To modify the gel time vary the amount of catalyst used in proportion to the resin.

NB To avoid wastage do not catalyse more resin than you can use in 15 minutes.

Mixing Polyester Resin

Mixing procedure

22 Pour the resin into a container.

23 Add the pigment to the resin and stir them together. NB Make sure that you have the required colour before going on to the next stage.

24 Add the catalyst to the mixture and stir them together. NB When the catalyst is added to the resin it tends to float on the surface and care must be taken to ensure that the two are properly mixed. Catalyst which is only partly mixed with the resin makes a fine streaky texture on the surface of the resin and it is not until this disappears and is replaced by a smooth surface that the catalyst is fully dispersed. From time to time hold the material to the light and check that the catalyst and resin are properly mixed.

25 NB When mixing polyester resin with catalyst and other materials it is very important to keep air out of the mix. Air which becomes trapped will mar the appearance of the finished work. The following points should be observed to prevent air getting into the mix.

a Use a tall container which, when holding the amount of resin to be used, is about two thirds full. Do not mix resin in a wide container as this would result in a larger surface area of resin which, when stirred, would be more likely to trap air (fig. 14).

b Always use a stirring implement that disturbs the surface of the resin as little as possible (fig. 14). For mixes under 2 oz it is a good idea to use a one to two-inch nail, or panel pin [brad], which can be thrown away after use.

c Stir gently and thoroughly.

14

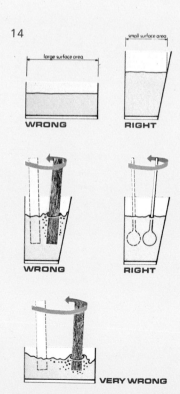

large surface area — WRONG

small surface area — RIGHT

WRONG   RIGHT

VERY WRONG

26 Removing air from the mixture

If a small quantity of air gets into catalysed resin, it can be removed by gently prodding out the individual air bubbles with the point of a panel pin. The best time to do this is immediately after the material has been catalysed and poured into the mould. Carry out this operation in a good light and work quickly, because as soon as the mix shows the least sign of gelling the task of air removing must be abandoned.

(Should an excessive amount of air get into uncatalysed resin, put the material to one side and let it stand until all the air has surfaced and dispersed, leaving the resin bubble free. This may take several hours or days, so be prepared to wait.)

Depth of the Casts

27 If the item to be made is going to be more than 7 mm [$\frac{1}{4}$ in] in depth it must be cast in more than one stage. As a general rule every additional 7 mm must be cast separately. Allow each stage to set properly before pouring the next one. Do not try to set too great a depth at any one time, otherwise the material will suffer a temperature rise that could destroy the mould.

Removing Casts from the Moulds

28 Usually the cast can be removed from the mould about two hours after it has been catalysed. Check whether or not the cast is ready for removal by gently flexing the mould. If the two materials part easily you may remove the casts by the following procedures.

29 Shallow moulds

i Flex the mould to expose the edge of the cast.
ii Grip the exposed edge and remove the cast from the mould.

30 Deep moulds

i Flex the mould to part it from the surface of the cast.
ii Gently bang the mould face downwards onto a surface until the cast falls in the mould.
iii Remove the cast from the mould by gripping the areas of the cast exposed by the cut-away section of the mould (p. 13 ‡ 9).

Finishing

15

31 Grinding

Occasionally it will be necessary to modify the shape of items that you have cast. Use a 120 grit silicon carbide paper (wet and dry), followed by a 500 grit paper. Always use water with the silicon carbide paper to keep down the dust and to assist the grinding process. An alternative method of grinding is to use a carborundum stone.

32 Polishing

Provided that the inside of the polythene moulds are scratch free, items cast in them should have a good finish on the surfaces which have been in contact with the mould. However, additional brilliance can be given to these surfaces by polishing them with a soft rag soaked in metal polish. The best results are obtained if this is done about a week after the cast has set, when it has had time to harden fully.

33 NB Surfaces treated with silicon carbide paper should be burnished with a wet cloth and abrasive powder and finished with a soft cloth soaked in metal polish.

34 Finishing the backs of the casts

When the casts are taken from the mould they have uneven backs. NB The back of a cast is the area which does not come into direct contact with the mould.

The unevenness may be removed by grinding the backs level (‡ 31 above). Do this by holding the sheet of silicon carbide paper on a flat surface with one hand and drawing the back of the cast to and fro across the wetted grit surface of the paper.

When the back of the cast has been reduced to a level surface it may be burnished with a wet cloth and abrasive powder and finished with a soft cloth soaked in metal polish.

Pouring Resin onto the backs of Casts

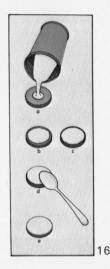

16

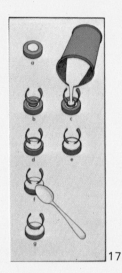

17

35 Many of the casts need to be given a backing of opaque polyester resin to reflect light through them and to provide a screen to prevent the metal mountings from showing through the casts.

36 Casts with embedded fittings (Rings)
i Place the cast, face downwards, on a level surface.

ii Estimate how much resin you will require to cover the back of the cast and measure the amount (p. 18 ‡ 20).

iii Pour the resin into a small container for mixing.

iv Mix in the opaque pigment (p. 16 ‡ 17).

v Pour 6 ml [6 cc] of the mixture into a second container.

vi Measure enough catalyst to catalyse the mixture in the second container (p. 18 ‡ 21).

vii Add the catalyst to the mixture and stir them together (p. 19 ‡ 24).

viii Pour a small amount of the catalysed material onto the centre of the back of the cast and allow it to gel (fig. 16a). NB The area covered should be large enough to hide the contact area of the ring when seen from the front of the cast.

ix Place the ring mounting, with the contact surface face downwards, at the centre of the back of the cast (fig. 16b). NB The tackiness of the gelled mixture will retain the ring mounting firmly in position.

x Catalyse the remaining mixture.

xi Carefully pour the catalysed mixture onto the back of the ring mounting through the break in the metal ring (fig. 16c). (The resin will flow over the ring mounting and onto the back of the ring.)

xii When the mixture has covered about $\frac{4}{5}$ of the surface, stop pouring and allow it to continue flowing under its own momentum (fig. 16d & e).

xiii Should the resin stop flowing short of the edge of the cast, use a blunt instrument, e.g. an old spoon, to encourage it to fill the remaining area (fig. 16f). NB Be careful not to allow the resin to flow down the sides of the cast.

xiv Examine the back of the cast to ensure that it is generously filled with resin. If more resin is required, pour it onto the back with great care, making sure that it does not flow over the sides.

xv Allow the resin to set. NB If the item is left untouched for a few days the tackiness will largely disappear from the surface, leaving a clean hard finish. The surface may then be given a light polish with a soft cloth containing metal polish. This will remove any lingering tackiness. If the ring mounting is made of brass or is gold plated, it is most likely lacquered, so take care not to remove the lacquer by polishing.

37 Casts without embedded fittings

i Place the cast face downwards on a level surface.

ii Estimate how much resin you will require to cover the back of the cast and measure the amount (p. 18 ‡ 20).

iii Pour the resin into a small container for mixing.

iv Mix in the opaque pigment (p. 16 ‡ 17).

v Measure the catalyst (p. 18 ‡ 21).

vi Add the catalyst to the mixture and stir them together (p. 19 ‡ 24).

vii Gently pour the mixture onto the back of the cast, beginning at the centre (fig. 17a).

viii When the mixture has covered about $\frac{4}{5}$ of the surface, stop pouring and allow it to continue flowing under its own momentum (fig. 17b and c).

ix Should the resin stop flowing short of the edge of the cast, use a blunt instrument, e.g. an old spoon, to encourage it to fill the remaining area. NB Be careful not to allow the resin to flow down the sides of the cast.

x Examine the back of the cast to ensure that it is generously filled with resin. If more resin is required pour it onto the back with great care, making sure that it does not flow over the sides.

xi Allow the resin to set. NB If the item is left untouched for a few days the tackiness will largely disappear from the surface, leaving a clean hard finish. The surface may then be given a light polish with a soft cloth containing metal polish. This will remove any lingering tackiness.

Translucent
Polyester Jewelry

Designing

When using natural stones to make jewelry the jeweller is limited in the designs he produces by the shape, size and colour of the stones available. The character of traditional jewelry is not altogether a matter of choice but rather a compromise between what the jeweller would like to make and what he is able to make with the material that is at his disposal. We can control the shape, size and colour of cast polyester in a way that would leave a jeweller of the past green with envy. Such good fortune should not be neglected. We would be very unwise simply to copy the style of conventional jewelry.

It is important to understand how to exploit translucent polyester resin to get the most interesting results.

It can be a material of unusual brilliance when it is used properly. Whichever way you choose to put it to work, you should take account of the play of light through the material. The following suggestions are given as a guide.

a Combine differently pigmented resins to play one colour off against another (fig. 26 p. 33).

b Use angled planes to reflect light through the cast (fig. 30 p. 37).

c Use an opaque background to reflect light through the face of a piece of jewelry (figs. 18, 19, 20 and 21 opposite).

These are just some of the ways in which translucent polyester can be exploited — others remain for you to discover.

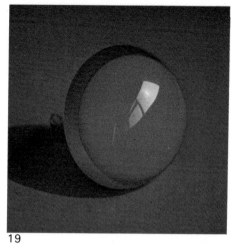

18

19

20

21

Project 1
Ring

Polyester Resin
Catalyst
Translucent pigment
Opaque pigment
Small polythene lid (mould)
2 small containers
1 Stirring implement
Ring mounting (or shank)

1 Estimate how much resin you will require to fill the mould and measure the amount (p. 18 ‡ 20).

2 Pour the resin into the first container for mixing.

3 Mix in the translucent pigment (p. 16 ‡ 16).

4 Measure the catalyst (p. 18 ‡ 21).

5 Add the catalyst to the mixture and stir them together (p. 19 ‡ 24).

6 Pour the mixture into the mould to the required depth.

7 Examine the mixture in the mould for air bubbles and remove any that may be present (p. 20 ‡ 26).

8 When the cast is set remove it from the mould (p. 20 ‡ 29).

22

9 Finish the back of the cast (p. 21
‡ 34).

10 Adjust the ring mountings to suit
the size of the finger of the
wearer.

11 Embed the ring mounting onto
the back of the cast (p. 22
‡ 36).

Project 2
Cuff-links

Polyester resin
Catalyst
Translucent pigment
Opaque pigment
2 small identical polythene lids
  (moulds)
3 small containers
Stirring implements
2 cuff-link mountings.

1 Estimate how much resin you
will require to fill the moulds and
measure the amount (p. 18
‡ 20).

2 Pour the resin into the first
container for mixing.

3 Mix in the translucent pigment
(p. 16 ‡ 16).

4 Measure the catalyst (p. 18
‡ 21)

5 Add the catalyst to the mixture
and stir them together (p. 19
‡ 24).

6 Pour the mixture into both
moulds to the required depth.

7 Examine the mixture in the
mould for air bubbles and
remove any that may be present
(p. 20 ‡ 26).

8 When the casts are set remove
them from the moulds (p. 20).

9 Finish the backs of the casts
(p. 21 ‡ 34).

10 Pour opaque resin onto the backs
of the casts (p. 23 ‡ 37).

11 Pour 6 ml [6 cc] of resin into
the third container and catalyse it.

12 Cover the contact surfaces of the
cuff-link mountings with the
resin and allow it to become
tacky.

13 Position the cuff-link mountings,
contact surface downward, at
the centre of the back of the
casts and apply a light downward
pressure. NB Make sure the
resin is tacky before the cuff-link
mountings are positioned or they
may slide over the surfaces of the
casts and set off-centre.

14 Allow the cuff-link mountings to
set in position.

23

Project 3
Magnifying Ring

Polyester resin
Catalyst
Translucent pigment
Opaque pigment
Measuring spoon (mould)
3 small containers
Stirring implements
Ring mounting (or shank)
Plasticine [plastiline]
Old water-colour paint brush
$1\frac{1}{2}$ inch nail

1 Press the bottom of the mould into the plasticine (to form a stable base) and check that it is level.

2 Estimate how much resin you will require to fill the mould and measure the amount (p. 18 ‡ 20).

3 Pour the resin into the first container for mixing.

4 Mix in the translucent pigment (p. 16 ‡ 16).

5 Measure the catalyst (p. 18 ‡ 21).

6 Add the catalyst to the mixture and stir them together (p. 19 ‡ 24).

7 Fill the mould with the mixture.

8 Examine the mixture in the mould for air bubbles and remove any that may be present (p. 20 ‡ 26).

9 When the cast is set remove it from the mould (p. 20 ‡ 29).

10 Finish the back of the cast (p. 21 ‡ 34).

11 Embed the cast face downward in the plasticine (to form a stable base) and check that it is level (fig. 24).

12 Pour 12 ml [12 cc] of resin into the second container.

13 Mix in the opaque pigment (p. 16 ‡ 18).

14 Pour $\frac{1}{3}$ (about 4 ml [4 cc]) of the mixture into the third container and catalyse it (3 drops of catalyst).

15 Use the paint brush to cover the contact surface of the ring mounting with the mixture and allow it to become tacky. NB Carry out this operation with the ring embedded in plasticine, contact surface facing upwards. The resin should be applied generously so that the contact surface does not show through the resin.

16 Immediately after completing ‡ 15, clean the paint brush with polyester cleaner.

17 Catalyse the remaining mixture.

24

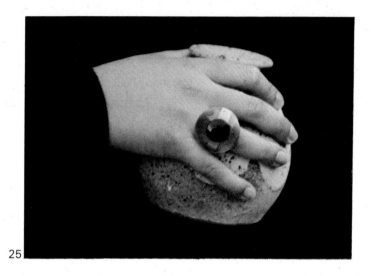

25

cast (fig. 24c) and gently bring the mounting down onto the resin (fig. 24d). NB This will cause the resin to spread and, if careful control is maintained, the result will be a perfect circle a little larger than the contact surface of the ring mounting.

18. Load a 1½ inch nail with the mixture and allow the mixture to drop slowly onto the centre of the cast (fig. 24a) covering a circular area equal to the area of the contact surface of the ring mounting (fig. 24b). NB This operation requires a steady hand and eye to deposit the resin exactly at the centre of the cast. You could practise depositing resin at the centre of a circle drawn on a sheet of paper first.

19 Hold the ring mounting with the contact surface face downwards exactly above the centre of the

20 Allow the resin to set. NB If the item is left untouched for a few days, the cured resin will generally lose its tackiness, leaving a clean finish. The surface may then be given a light polish with a soft cloth containing metal polish, which will remove any lingering tackiness. If the ring mounting is made of brass or is gold plated, it is most likely lacquered, so take care not to remove the lacquer by rubbing the metal with the impregnated cloth.

Project 4
Brooch

Polyester Resin
Catalyst
2 translucent pigments
Polythene lid (mould)
3 small containers
Stirring implements
Brooch pin

1 Estimate how much resin you will require to fill the mould and measure the amount (p. 18 ‡ 20).

2 Pour the resin into the first container for mixing.

3 Mix in the translucent pigment (p. 16 ‡ 16).

4 Measure the catalyst (p. 18 ‡ 21).

5 Add the catalyst to the mixture and stir them together (p. 19 ‡ 24).

6 Pour the mixture into the mould so that it reaches a level just short of the rim.

7 Examine the mixture in the mould for air bubbles and remove any that may be present (p. 20 ‡ 26).

8 Allow the mixture to set.

9 Pour a small quantity of resin into the second container.

10 Mix in the second translucent pigment.

11 Catalyse the mixture and pour it on top of the first layer of polyester, filling the mould to the rim.

12 Examine the mixture in the mould for air bubbles and remove any that may be present.

13 Allow the mixture to set.

14 Remove the cast from the mould (p. 20 ‡ 29).

15 Finish the back of the cast (p. 21 ‡ 34).

16 Pour opaque resin onto the back of the cast (p. 23 ‡ 37).

17 Pour 6 ml [6 cc] of resin into the third container and catalyse it.

26

18 Cover the face of the brooch pin
mounting with the resin and
allow it to become tacky.

19 Position the brooch pin, contact
surface downwards, at the centre
of the back of the cast and apply
a light downward pressure to the
pin. NB Make sure the resin
is tacky before the brooch pin is
positioned or it may slide over
the surface of the cast and set
off-centre.

20 Allow the brooch pin to set in
position.

Project 5
Scarf Ring
Cuff-links

Polyester resin
Catalyst
Translucent pigment
Opaque pigment
2 small identical polythene lids
   (moulds)
1 medium polythene lid (mould)
4 small containers
Stirring implements
A small metal file
2 cuff-link mountings
A 6 mm [$\frac{1}{4}$ in] length of 19 mm
   [$\frac{3}{4}$ in] dia. brass or copper tube
   (ring)

1 Estimate how much resin you
will require to fill the moulds and
measure the amount (p. 18
‡ 20).

2 Pour the resin into the first
container for mixing.

3 Mix in the translucent pigment
(p. 16 ‡ 16).

4 Measure the catalyst (p. 18
‡ 21).

5 Add the catalyst to the mixture
and stir them together (p. 19
‡ 24).

6 Pour the mixture into each of the
three moulds to the required
depth.

7 Examine the mixture in the
moulds for air bubbles and
remove any that may be present
(p. 20 ‡ 26).

8 When the casts are set remove
them from the moulds (p. 20
‡ 29).

9 Finish off the backs of the casts
(p. 21 ‡ 34).

Scarf Ring

10 File a small flat area on the
surface of the brass or copper
ring. NB This is to prevent the
ring from moving when it is
placed onto the back of the cast.

11 Carry out operations i–ix on
p. 22, ‡ 37.

12 Before the resin has time to gel,
gently embed the brass or
copper ring, flat downwards,
in the centre of the resin.

13 Allow the resin to set. NB If the
scarf ring is left untouched for
a few days the tackiness will
largely disappear from the back,
leaving a clean hard finish. The
surface may then be given a light
polish with a soft cloth containing metal polish. This will
remove any lingering tackiness.

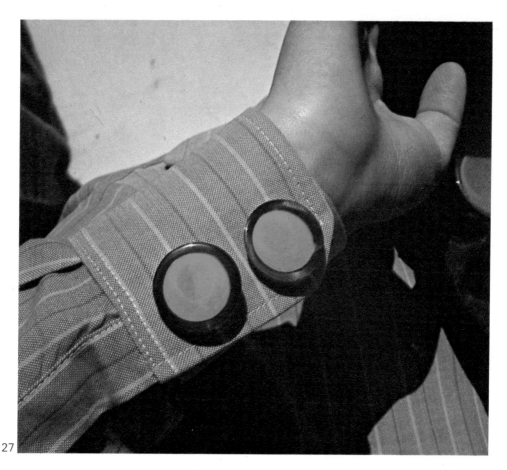

27

Cuff-links

14 Pour opaque resin onto the backs of the two remaining casts (p. 23 ‡ 37).

15 Pour 6 ml [6 cc] of resin into the fourth container and catalyse it.

16 Cover the contact surfaces of the cuff-links with the resin and allow it to become tacky.

17 Position the cuff link mountings, contact surface downwards, at the centres of the back of the casts and apply a light downward pressure to them. NB Make sure the resin is tacky before the cuff-link mountings are positioned or they may slide over the surfaces of the casts and set off-centre.

18 Allow the cuff-link mountings to set in position.

Project 6
Ring

28

Polyester resin
Catalyst
Translucent pigment
Opaque pigment
Small polythene bottle (mould)
   (fig. 29)
Three small containers
Stirring implements
Ring mounting (or shank)

1 Prepare the polythene bottle for use as a mould (p. 13 ‡ 9).

2 Estimate how much resin you will require to fill the mould and measure the amount (p. 18 ‡ 20).

3 Pour the resin into the first container for mixing.

4 Mix in the translucent pigment (p. 16 ‡ 16).

5 Measure the catalyst (p. 18 ‡ 21).

6 Add the catalyst to the mixture and stir them together (p. 19 ‡ 24).

7 Pour the mixture into the mould, covering the raised centre by about 3 mm [$\frac{1}{8}$ in].

8 Examine the mixture in the mould for air bubbles and remove any that may be present (p. 20 ‡ 26).

9 When the cast is set remove it from the mould (p. 20 ‡ 30).

10 Finish the back of the cast (p. 21 ‡ 34).

11 Pour a small quantity of resin into the second container.

12 Mix in the opaque pigment (p. 16 ‡ 18).

13 Pour a third of the mixture into the third container and catalyse it.

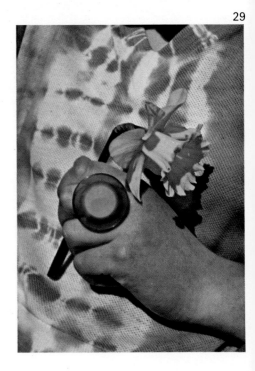

29

30

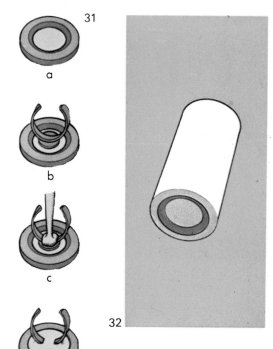

31

a

b

c

d

32

14 Pour enough of the mixture into the recessed surface of the cast just to cover it (fig. 31a).

15 Allow the mixture to set.

16 Adjust the ring mounting to suit the finger size of the wearer.

17 Place the ring mounting with its contact surface downwards at the centre of the recessed surface and apply a light downwards pressure (fig. 31b). NB The tacky surface will retain the ring in position and prevent it moving.

18 Catalyse the remaining opaque mixture.

19 Pour the mixture through the gap in the ring mounting, onto the recess at the back of the cast (fig. 31c), filling the recess level with the top of the cast (fig. 31d).

20 Allow the mixture to set. NB If the item is left untouched for a few days, the cured mixture will generally lose its tackiness, leaving a clean finish. The surface may then be given a light polish with a soft cloth containing metal polish, which will remove any lingering tackiness. If the ring mounting is made of brass or is gold plated, it is most likely lacquered, so take care not to remove the lacquer with the impregnated cloth.

Project 7
Cylindrical Ring

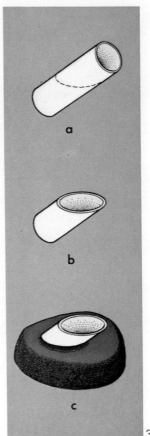

a

b

c

33

Polyester resin
Catalyst
Translucent pigment
Opaque pigment
Narrow polythene pill-box
2 containers
Stirring implements
Ring mounting (or shank)
Plasticine [plastiline]

1 Prepare the polythene pill-box for use as a mould, i.e. cut the pill-box to the required length with a diagonal cut (fig. 33a).

2 Press the mould into the plasticine so that the open end of the mould is uppermost and horizontal (fig. 33c).

3 Estimate how much resin you will require to fill the mould and measure the amount (p. 18 ‡ 20).

4 Pour the resin into the first container for mixing.

5 Mix in the translucent pigment (p. 16 ‡ 16).

6 Measure the catalyst (p. 18 ‡ 21).

7 Add the catalyst to the mixture and stir them together (p. 19 ‡ 24).

8 Pour the mixture into the mould filling it to the rim.

34

9 Examine the mixture in the mould
for air bubbles and remove any
that may be present (p. 20
‡ 26).

10 When the cast is set remove it
from the mould (p. 20 ‡ 30).

11 Finish the back of the cast
(p. 21 ‡ 34).

12 Press the cast into the plasticine
so that the diagonal back of the
cast is horizontal (similar to
operation 2) (fig. 33c).

13 Adjust the ring mounting to suit
the size of the wearer's finger.

14 Embed the ring mounting onto
the back of the cast (p. 22
‡ 36).

Inlaid Jewelry

35

You may have noticed that when resin is poured into a mould it tends to ride a little way up the sides (fig. 38b). The resin will also ride up the sides of any object that is placed in the mould (provided the object stands higher than the surface of the resin). If we put a number of small casts in a polythene lid and then pour two layers of different coloured resins around them, we can produce a larger cast which, when levelled, will give an inlaid effect.

As this new method of employing polyester resin has not been fully exploited, there still remains a lot of scope for designing and making original pieces of jewelry. The items illustrated show some of the ways in which the inlaid technique can be used. See how many original variations you can produce after doing the projects.

36

37

Project 8
Medallion

Polyester resin
Catalyst
Translucent pigment
Opaque pigment
Large polythene lid (mould)
2 polythene nozzles (from liquid
detergent bottles)
About 8 containers
Stirring implements
Ribbon

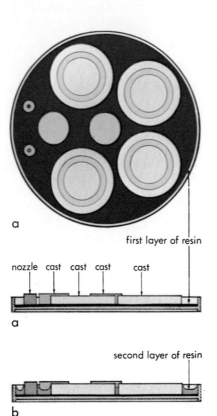

a

first layer of resin

nozzle cast cast cast   cast

a

second layer of resin

b

third layer of resin

c

hole

d

38

1 Make 4 medium and 2 small-
sized casts which, when placed
in the mould, will stand higher
than the resin that is to be poured
into them. NB Follow the
instructions given in projects
Nos 1 and 6 (pp. 26 and 36)
to make the casts but leave them
unbacked.

2 Prepare the polythene nozzles,
i.e. cut them to length (they
should be long enough to stand
well proud of the finished cast
to make them easy to remove).

3 Measure 6 ml [6 cc] of resin
(p. 18 ‡ 20).

4 Pour the resin into a container
for mixing.

5 Measure the catalyst (p. 18
‡ 21).

6 Add the catalyst to the resin and
stir them together (p. 19 ‡ 24).

7 Pour enough of the catalysed
resin into the mould just to cover
the bottom (first layer – shown
purple in fig. 38a).

8 When the resin becomes tacky
   arrange the casts and the two
   nozzles in the mould (fig. 38a).
   NB When removed from the cast
   the nozzles will leave two holes
   for the ribbon.

9 Estimate how much resin you
   will require nearly to fill the
   remaining area in the mould and
   measure the amount (p. 18
   ‡ 20).

10 Pour the resin into a container
   for mixing.

11 Mix in the translucent pigment
   (p. 16 ‡ 16).

12 Measure the catalyst (p. 18
   ‡ 21).

39

13 Add the catalyst to the mixture
   and stir them together (p. 19
   ‡ 24).

14 Pour the mixture into the mould
   (second layer – shown brown in
   fig. 38b). NB The level reached
   by the resin as it rides up the
   sides of the mould and casts
   should be just short of the tops
   of both (fig. 38b).

15 Examine the mixture in the
   mould for air bubbles and
   remove any that may be present
   (p. 20 ‡ 26).

16 Allow the resin to set.

17 Pour enough resin just to cover
   the surface of the cast, into a
   container for mixing.

18 Mix in the opaque pigment
   p. 16 ‡ 18).

19 Catalyse the mixture.

20 Pour the mixture onto the
   previous layer of material in the
   mould (third layer – shown
   black in fig. 38c). Use no more
   material than is necessary just
   to cover the surface.

21 Allow the mixture to set.

22 Remove the cast from the mould
   (p. 20 ‡ 29).

23 Remove the polythene nozzles
   from the cast.

24 Level the inlaid casts to the
   surface of the surrounding
   material (p. 21 ‡ 31) (fig. 38d).

25 Finish the cast (p. 21 ‡ 33).

26 Thread the ribbon through the
   two holes, cut it to length and
   sew the two ends together.

Project 9
Pendant

Polyester resin
Catalyst
Translucent pigment
Opaque pigment
Large polythene lid (mould)
Small polythene pill-box top
About 6 containers
Stirring implements
2 jump rings
Chain and bolt ring [spring ring]
Old water-colour paint brush

1 Make two casts (casts A) similar to the one described in project 6 (p. 36) but leave them unbacked.

2 Make two additional casts (casts B) similar to the one described in project 3 (p. 30) but leave them unbacked.

3 Measure 6 ml [6 cc] of resin (p. 18 ‡ 20).

4 Pour the resin into a container for mixing.

5 Measure the catalyst (p. 18 ‡ 21).

6 Add the catalyst to the resin and stir them together (p. 19 ‡ 24).

7 Pour enough of the catalysed resin into the mould just to cover the bottom.

8 When the resin becomes tacky, arrange casts A and the polythene pill-box top in the mould

(similar to fig. 38a). NB When removed from the cast the pill-box will leave a hole for the jump rings.

Note
The next four operations must be performed with particular care to avoid introducing air into the mixture. As we shall be dealing with an opaque resin it will be impossible to see, and thus remove, any air bubbles that may be below the surface.

9 Estimate how much resin you will require nearly to fill the remaining area in the mould and measure the amount (p. 18 ‡ 20).

10 Pour the resin into a container for mixing.

11 Mix in the opaque pigment (p. 16 ‡ 17).

12 Measure the catalyst (p. 18 ‡ 21).

13 Add the catalyst to the mixture and stir them together (p. 19 ‡ 24).

14 Pour the mixture into the mould. NB The level reached by the resin as it rides up the sides of the mould and casts should be just short of the tops of both (similar to fig. 38b).

15 Examine the mixture in the mould for air bubbles and remove any that may be present (p. 20 ‡ 26). NB As the mixture is opaque it will only be possible to remove air bubbles from the surface of the mixture.

40

16 Allow the resin to set.

17 Pour enough resin just to cover the surface of the cast into a container for mixing.

18 Mix in the translucent pigment (p. 16 ‡ 16).

19 Catalyse the mixture.

20 Pour the mixture on to the previous layer of material in the mould. Use no more of the mixture than is necessary just to cover the surface (similar to fig. 38c).

21 Allow the mixture to set.

22 Remove the cast from the mould (p. 20 ‡ 29).

23 Remove the pill-box top from the cast.

24 Level the inlaid casts to the surface of the surrounding material (p. 21 ‡ 31) (similar to fig. 38d).

25 Finish the cast (p. 21 ‡ 33).

26 Place the cast on a level surface, face upwards.

27 Pour 6 ml [6 cc] of clear resin into a container and catalyse it.

28 Using the paint brush, cover the backs of casts B with the resin.

29 When the resin just begins to gel, very carefully place casts B with their backs facing downwards, at the centres of the two inlaid casts (casts A) and apply a light downward pressure.

30 Wait until the applied casts are stuck firmly (a few minutes) and carefully remove any surplus resin from the surfaces of the casts with a soft cloth containing metal polish. NB Timing is very important in carrying out this operation. The surplus resin must be removed before it has prop-erly set and when the applied casts are firm enough not to be disturbed by the cleaning action.

31 Immediately after completing 30, clean the brush with polyester cleaner.

32 Allow the casts to set in position.

33 Attach the jump rings and thread the chain through them.

34 Attach the bolt ring [spring ring] to the chain.

Embedded Jewelry

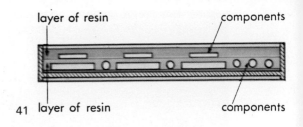

layer of resin   components

41 layer of resin   components

The following examples of jewelry can be made with a wide variety of components, many of which can be bought cheaply from haberdashers or rescued from workshop waste. Most small objects which are dry and reasonably inert chemically, can be used, e.g. dried peas, pieces of broken jewelry, old buttons, etc.

The type of components that you can lay your hands on should themselves provide the key to the design of your jewelry, so it is useful to have a fair collection of bits and pieces from which to choose. By going over your store of knick-knacks and paying a visit to the garage or garden shed, you should be able to establish an adequate collection. Later it can be extended with the booty obtained from sorties to the local surplus store. Second hand electrical components shops have a wealth of cheap washers, fasteners and other mysterious and sometimes beautiful objects that are for sale in job lots.

At this rate it will not be long before you will have to start organizing your treasure trove. One cheap and effective way to do this is to make use of those ever present egg cartons. Placed side by side in a drawer or cardboard box (fig. 43) they provide a good method of 'filing' your components, making it easy to select the pieces that you require and also to see exactly what you have in stock at any one time.

Designing
When selecting components for your design, everything of suitable size should be considered, nothing is too humble for inclusion. Some of the most unlikely objects can produce beautiful and remarkable effects if placed together imaginatively.

The easiest way to design a piece of jewelry is to arrange and re-arrange different components from your collection until you have what you consider to be an attractive design. Do this in an identical mould to the one in which the piece is to be cast.

42

Whilst working on a design it should
be remembered that you are actually
looking at the back of what will be
the finished piece of jewelry. At first
you may find difficulty in visualizing
how your arrangement will look
when removed from the mould and
seen from the other side but, with
practice, you will find that visualizing in this way becomes second
nature.

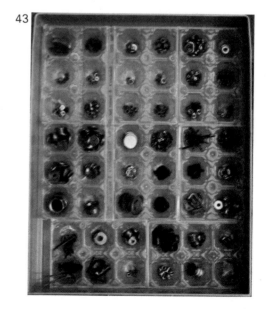

43

By pouring the resin into the mould
in stages, the components can be
sandwiched between each layer of
resin (fig. 41), thus giving the
finished item a three dimensional
appearance. Arranging the components in this way can give extra
richness to a piece of jewelry.

47

Project 10
Brooch

Polyester resin
Catalyst
Opaque polyester pigment
2 identical polythene lids (moulds)
5 containers
Brooch pin

Optional:
Sequins
Washers
Spring washers
Cake cachous [ball-bearings would do]
Glitter
Metallic paper

1 Position some of the decorative components in the bottom of the first mould to form an attractive arrangement. (The sequins are best positioned with the point of a sharpened wax crayon as they are difficult to manipulate with the fingers.) NB Bear in mind that some of the components are to be set on one level and the remainder on a second level, to give a three dimensional appearance. Make a mental note of which components are to appear on each level. Remember that your arrangements will be seen from the other side when the brooch is taken from the mould (p. 47).

2 Measure 6 ml [6 cc] of resin (p. 18 ‡ 20).

3 Pour the resin into the first container for mixing.

4 Measure the catalyst (p. 18 ‡ 21).

5 Add the catalyst to the resin and stir them together (p. 19 ‡ 24).

6 Pour enough of the catalysed resin into the second mould just to cover the bottom.

7 Immediately the resin has gelled, transfer the decorative components belonging to the first level of your arrangement, one at a time, onto the surface of the gelled resin. NB Care should be taken to position the components in exactly the same place that they occupied in the first mould, because due to the tackiness of the resin it is difficult to re-position them once they are in place. It is important that the resin should be tacky before trying to arrange the components, otherwise they will float about.

8 Pour enough resin to fill the mould to a depth of 3 mm [$\frac{1}{8}$ in] into the second container and catalyse it.

9 Pour it into the second mould to a depth of 3 mm [$\frac{1}{8}$ in].

10 Examine the resin in the mould for air bubbles and remove any that may be present (p. 20 ‡ 26).

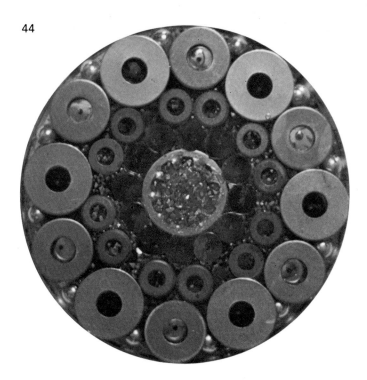

11 Immediately after the resin has gelled, position the remaining components, e.g. sequins, in the centres of the washers and the metallic paper in the centre of the spring washer.

12 Pour enough resin to fill the mould completely into the third container and catalyse it.

13 Fill the second mould to the brim.

14 When the cast is set, remove it from the mould (p. 20 ‡ 29).

15 Finish the back of the cast (p. 21 ‡ 34).

16 Pour opaque resin onto the back of the cast (p. 23 ‡ 37).

17 Cover the contact surface of the brooch pin with clear catalysed resin and allow it to become tacky.

18 Position the brooch pin, contact surface downwards, at the centre of the back of the cast and apply a light downward pressure. NB Make sure the resin is tacky before the brooch pin is positioned or it may slide over the surface of the cast and set off-centre.

19 Allow the brooch pin to set in position.

Project 11
Scarf Ring Cuff-links

Polyester resin
Catalyst
Black opaque pigment
7 containers
4 measuring spoons (moulds)
Stirring implements
File
A 6 mm [$\frac{1}{4}$ in] length of 19 mm [$\frac{3}{4}$ in]
    dia. brass or copper tube (ring)
2 cuff-link mountings
Plasticine [plastiline]

Optional:
Sequins—2 colours
3 spring washers
3 artificial pearls
3 circles of silver metallic paper
    (half the diameter of the mould)

1 Press the bottoms of the moulds into individual pieces of plasticine and check that they are level.

2 Arrange the decorative components in one of the moulds to form a satisfactory design, which should then be put to one side to serve as a guide. NB If you have difficulty keeping the components in position on the steep surface of the mould, stick them into place temporarily with cow gum [rubber cement] or other suitable adhesive. Make sure that you have enough components to embed in all three casts.

3 Measure 6 ml [6 cc] of resin (p. 18 ‡ 20).

4 Pour the resin into the first container for mixing.

5 Measure the catalyst (p. 18 ‡ 21).

6 Add the catalyst to the resin and stir them together (p. 19 ‡ 24).

7 Cover the inside surface of the remaining three moulds with the resin by running a small amount around the inside. NB Turn the moulds upside down on a piece of newspaper to allow the surplus resin to drain out.

8 When the resin has set repeat ‡ 3, 4, 5, 6 and 7.

9 Immediately the second application of resin has gelled, place the decorative components, one at a time, in their correct position in each of the three moulds. NB Care should be taken to arrange the components in their exact positions because, due to the tackiness of the resin, it is difficult to re-position them once they are in place. It is important that the resin should be tacky before trying to arrange the components, otherwise they will slide down the sides of the moulds.

10 Estimate how much resin you will require to fill the moulds to a depth of 6 mm [$\frac{1}{4}$ in] and measure the amount (p. 18 ‡ 20).

11 Pour the resin into the third container and catalyse it.

12 Pour the catalysed resin into each of the moulds to a depth of 6 mm [$\frac{1}{4}$ in].

45

13 Examine the resin in the moulds for air bubbles and remove any that may be present (p. 20 ‡ 26).

14 When the resin is set, place one circle of paper, silver side downwards, on to the surface of the resin in each of the moulds. Apply a light pressure to the back of the paper so that it makes a good contact.

15 Pour 6 ml [6 cc] of resin into the fourth container and catalyse it.

16 Pour the resin into each of the three moulds to cover the circles of paper.

17 Examine the resin in the moulds for air bubbles and remove any that may be present.

18 When the casts are set, remove them from the moulds (p. 20 ‡ 29).

19 Finish the backs of the casts (p. 21 ‡ 34).

Scarf Ring

20 File a small flat area on the outer surface of the brass or copper ring. NB This is to prevent the ring from moving when it is placed on the back of the cast.

21 Press one of the casts face downwards into plasticine (to form a stable base) and check that it is level.

22 Carry out operations ii–ix on p. 23 ‡ 37.

23 Before the resin has time to gel, gently embed the brass or copper ring flat filed surface downwards in the centre of the resin.

24 Allow the resin to set. NB If the scarf ring is left untouched for a few days the tackiness will largely disappear from the back, leaving a clear hard finish. The surface may then be given a light polish with a soft cloth containing metal polish. This will remove any lingering tackiness.

46

Project 12
Pendant

Cuff-links

25 Press the two remaining casts face downwards into plasticine (to form a stable base) and check that they are level.

26 Pour opaque resin onto the back of the casts (p. 23 ‡ 37).

27 Pour 6 ml [6 cc] of resin into the seventh container and catalyse it.

28 Cover the contact surface of the cuff-link mountings with the resin and allow it to become tacky.

29 Position the cuff-link mountings, with the contact surface face downwards, at the centre of the back of the casts and apply a light pressure to them. NB Make sure the resin is tacky before the cuff-link mountings are positioned or they may slide over the surfaces of the casts and set off-centre.

30 Allow the cuff-link mountings to set in position.

Polyester resin
Catalyst
Opaque polyester pigment
2 identical polythene lids (moulds)
3 containers
Stirring implements
Wire
12 Swedish or horse nails
Leather thong

Optional:
Sequins and Glitter
Washers
Circle of metallic paper
Small beads

1 Bend the wire to form the hook shown in fig. 47.

2 Position the decorative components in the bottom of the first mould to form an attractive arrangement. (The sequins are best positioned with the point of a sharpened wax crayon as they are difficult to manipulate with the fingers.) NB Your arrangement will be seen from the other side when taken from the mould (p. 47).

47b

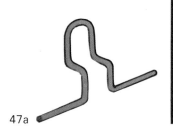

47a

3 Measure 6 ml [6 cc] of resin (p. 18 ‡ 20).

4 Pour the resin into the first container for mixing.

5 Measure the catalyst (p. 18 ‡ 21).

6 Add the catalyst to the resin and stir them together (p. 19 ‡ 24).

7 Pour enough of the resin in the mould just to cover the bottom.

8 Examine the resin in the mould for air bubbles and remove any that may be present (p. 20 ‡ 26).

9 Immediately the resin has gelled, transfer the decorative components, one at a time, to the second mould. NB Care should be taken to arrange the components in exactly the same position that they occupied in the first mould, because, due to the tackiness of the resin, it is difficult to re-position the components once they are in place. It is important that the resin is tacky before trying to arrange the components, otherwise they will float about.

10 Arrange half the Swedish nails at equal distances around the mould with their points just touching at the centre of the mould.

11 Lightly sprinkle the glitter onto the backs of the components allowing it to fall between them.

12 Pour enough resin to cover the backs of the components to a depth of 3 mm [$\frac{1}{8}$ in], into the second container and catalyse it.

13 Pour the resin into the mould over the backs of the components to a depth of 3 mm [$\frac{1}{8}$ in].

14 Examine the resin in the mould for air bubbles and remove any that may be present.

15 Immediately the resin has gelled position the circle of metallic paper at the centre of the mould and apply a light downward pressure to it.

48

16 Arrange the remaining nails at equal distances between the nails that are already embedded in the mould. NB The points of the nails should just touch at the centre of the back of the metallic paper.

17 Position the hook at the back of the mould.
i The hook should be placed with the loop astride one of the nails already embedded in the resin and with each tail under one of the loose nails at either side.

ii The hook should be placed as high as possible at the back of the mould but it should not show from the front of the finished cast, p. 53, fig. 47b

Note

The next four operations must be performed with particular care to avoid introducing air into the mixture. As we shall be dealing with an opaque resin it will be impossible to see, and thus remove, any air bubbles that may be below the surface.

18 Pour enough resin into the third container to fill the mould.

19 Mix in the opaque pigment (p. 16 ‡ 18).

20 Catalyse the mixture.

21 Pour the mixture into the mould, filling it to the rim.

22 Examine the mixture in the mould for air bubbles and remove any that may be present. NB As the mixture is opaque it will only be possible to remove air bubbles from the surface of the mixture.

23 Allow the mixture to set. NB If the cast is left untouched for a few days the cured polyester will generally lose its tackiness, leaving a clean finish. The surface may then be given a light polish with a soft cloth, which will remove any lingering tackiness.

24 Remove the cast from the mould (p. 20 ‡ 29).

25 Thread the leather thong through the hook and tie the two ends together.

Project 13
Magnifying Ring

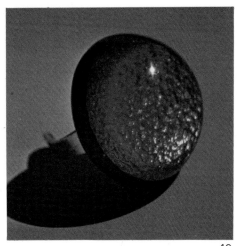

49

Polyester resin
Catalyst
2 translucent polyester pigments
Opaque polyester pigment
Measuring spoon (mould)
4 containers
Stirring implements
Heavily embossed metallic paper or
   plastic sheet
Ring mounting (or shank)
Plasticine [plastiline]
Old water-colour paint brush

1 Cut a circle from the embossed
   metallic sheet $\frac{4}{5}$ the diameter of
   the rim of the mould.

2 Press the bottom of the mould
   into the plasticine (to form a
   stable base) and check that it is
   level.

3 Estimate how much resin you
   will require to fill the mould
   $\frac{3}{4}$ full and measure the amount
   (p. 18 ‡ 20).

4 Pour the resin into the first container for mixing.

5 Mix in the first translucent
   pigment (p. 16 ‡ 16).

6 Measure the catalyst (p. 18
   ‡ 21).

7 Add the catalyst to the mixture
   and stir them together (p. 19
   ‡ 24).

8 Pour the mixture into the mould
   so that it is $\frac{3}{4}$ full.

9 Examine the mixture in the mould
   for air bubbles and remove any
   that may be present (p. 20
   ‡ 26).

10 Allow the mixture to gel.

11 Pour 6 ml [6 cc] of resin into the
   second container for mixing.

12 Mix in the second translucent
   pigment and catalyse the mixture.

50

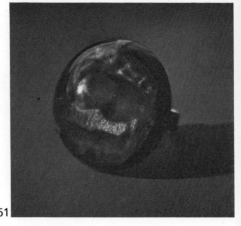

51

Note
The next five operations must be carried out smartly before the mixture has time to gel.

13 Pour some of the mixture into the mould so that it is $\frac{4}{5}$ full.

14 Examine the mixture in the mould for air bubbles and remove any that may be present.

15 Using the paint brush, immediately cover the face of the metallic sheet with the mixture remaining in the second container. NB This will reduce the possibility of trapping air under the sheet when it is placed in the mould.

16 Immediately position the metallic sheet, resin side down, in the centre of the mould and apply a light downward pressure. NB The application of pressure should force out any air that may be trapped under the sheet.

17 Clean the brush with polyester cleaner.

18 Allow the metallic sheet to set in position.

19 Pour enough resin into the third container to fill the mould and catalyse it.

20 Pour the catalysed resin into the mould, filling it to the rim.

21 Examine the resin for air bubbles and remove any that may be present.

22 Allow the resin to set.

23 Remove the cast from the mould (p. 20 ‡ 29).

24 Finish the back of the cast (p. 21 ‡ 34).

25 Press the cast face downward in the plasticine (to form a stable base) and check that it is level.

26 Embed the ring mounting in the back of the cast (p. 22 ‡ 36).

Project 14
Brooch

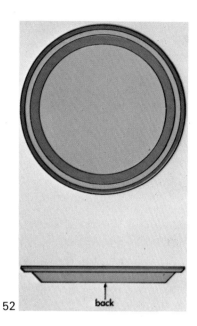

52   back

Polyester resin
Catalyst
3 translucent pigments
Measuring spoon (mould)
Polythene lid (preferably with a
    shaped profile) (mould)
7 containers
Stirring implements
Embossed metallic paper or plastic
    sheet
Brooch mounting
Plasticine [plastiline]
Old water-colour paint brush

1 Carry out operations 1–24 of
   project 13 on page 55 using the
   measuring spoon as the mould to
   make the first cast (cast A).

2 Estimate how much resin you
   will require to fill the polythene
   lid and measure the amount
   (p. 18 ‡ 20).

3 Pour the resin into the fifth
   container.

4 Mix in the third translucent
   pigment.

5 Measure the catalyst (p. 18
   ‡ 21).

6 Add the catalyst to the mixture
   and stir them together (p. 19
   ‡ 24).

7 Pour the mixture into the lid.

8 Examine the mixture in the lid for
   air bubbles and remove any that
   may be present (p. 20 ‡ 26).

9 Allow the mixture to set.

10 Remove the cast (cast B) from
   the mould (p. 20 ‡ 29).

11 Finish the back of the cast
   (p. 21 ‡ 34).

53

whether or not the resin has gelled, check the remaining resin in the sixth container. The resin must be tacky when the two casts are brought together or they may slide over one another and off-centre.

15 Immediately after completing 13 clean the brush with polyester cleaner.

16 Allow the casts to set together.

17 Press the assembled casts, face downward, in plasticine (to form a stable base).

18 Pour 6 ml [6 cc] of resin into the seventh container and catalyse it.

19 Using the brush, 'paint' the resin onto the back of cast B to within 3 mm [$\frac{1}{8}$ in] of the outer diameter of cast A. NB The resin should not be so thick that it will spread beyond the outer diameter of cast A when ‡ 20 is carried out.

20 When the resin just begins to gel, very carefully place the brooch pin with its contact surface face downward at the centre of the back of cast B and apply a light downward pressure. NB Make sure the resin is tacky before the brooch pin is positioned or it may slide over the surface of the cast and set off-centre.

21 Immediately after completing ‡ 19 clean the brush with polyester cleaner.

22 Allow the brooch pin to set in position.

Note
If your cast has shaped sides, use it the way up shown in fig. 52, i.e., the side which has just been finished (‡ 11 above) becomes the front. If you have a cast which has straight sides, it is unimportant which side is treated as the back.

12 Pour 6 ml [6 cc] of resin into the sixth container and catalyse it.

13 Using the brush 'paint' the catalysed resin onto the back of cast A to within 3 mm [$\frac{1}{8}$ in] of its outer diameter. NB The resin should not be so thick that it will be forced out from under cast A when ‡ 14 is carried out.

14 When the resin begins to gel, very carefully place cast A with its back facing downwards at the centre of the front of cast B and apply a light downward pressure. NB To determine

58

Project 15
Kinetic Ring

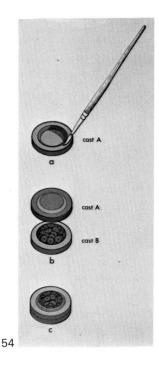

54

Polyester resin
Catalyst
Translucent pigment
Opaque pigment
2 identical small polythene bottles
    for moulds (p. 37 fig. 32).
Sequins (or other small components)
Stirring implements
Ring mounting (or shank)
Old water-colour paint brush

1 Prepare the polythene bottles for use as moulds (p. 13 ‡ 9).

2 Estimate how much resin you will require to fill the moulds and measure the amount (p. 18 ‡ 20).

3 Pour the resin into a container for mixing.

4 Mix in the translucent pigment (p. 16 ‡ 16).

5 Measure the catalyst (p. 18 ‡ 21).

6 Add the catalyst to the mixture and stir them together (p. 19 ‡ 24).

7 Pour the mixture into the two moulds, covering the raised centre by about 3 mm [$\frac{1}{8}$ in].

8 Examine the mixture in the moulds for air bubbles and remove any that may be present (p. 20 ‡ 26).

9 When the casts are set, remove them from the moulds (p. 20 ‡ 30).

10 Finish the backs of the casts (p. 21 ‡ 34).

11 Ensure that the front surfaces of both casts are perfectly flat by lightly grinding them with 500 grit paper (p. 21 ‡ 31).

12 Finish the front surfaces with a soft rag soaked in metal polish.

59

55

17 Position cast A over cast B, making sure they are properly aligned and press firmly in place (fig. 54b and c). NB Do not disturb the sequins.

18 Clean the paint brush with the polyester cleaner.

19 Allow the two casts to set together. NB Under no circumstances must they be moved until they are properly set, otherwise the sequins will be disturbed and may stick to the unset resin.

20 With 500 grit silicon carbide paper, remove any surplus resin from the outsides of the casts and burnish the treated surface with a wet cloth and abrasive powder. Finish the surface with a soft rag soaked in metal polish.

21 Adjust the ring mounting to suit the size of the wearer's finger.

22 Place the casts on a level surface with the back of cast B face upwards.

23 Embed the ring mounting in the back of cast B (p. 22 ‡ 36).

13 Place one of the casts (cast B) back downwards on a level surface.

14 Pile the sequins into the recess of cast B (fig. 54b).
NB When the two casts are placed face to face, a hollow centre is formed. Enough sequins should be piled into the recess of cast B to fill $\frac{3}{4}$ of the hollow centre. This will ensure that when the ring is agitated the sequins will move about freely, thus creating the kinetic effect.

15 Pour 6 ml [6 cc] of resin into a container and catalyse it.

16 Place cast A back downwards on a flat surface and with the brush 'paint' the resin on the front surface of the cast (fig. 54a) NB Care must be taken not to get the resin in the recess.

Project 16
Cuff-links

56

Polyester resin
Catalyst
Translucent pigment
Opaque pigment
2 measuring spoons (moulds)
4 containers
Stirring implements
2 black spring washers
2 cuff-link mountings
Plasticine [plastiline]

1 Press the bottoms of the moulds into individual pieces of plasticine (to form a stable base) and check that they are level.

2 Estimate how much resin you will require to fill the moulds and measure the amount (p. 18 ‡ 20).

3 Pour the resin into the first container for mixing.

4 Mix in the translucent pigment (p. 16 ‡ 16).

5 Into the second container pour enough of the mixture to fill the moulds three-quarters full.

6 Measure enough catalyst to catalyse the mixture in the second container (p. 18 ‡ 21).

7 Add the catalyst to the mixture in the second container and stir them together (p. 19 ‡ 24).

8 Three-quarters fill the moulds from the second container.

9 Examine the mixture in the moulds for air bubbles and remove any that may be present (p. 20 ‡ 26).

10 Allow the mixture to gel.

11 Carefully place the washers centrally on top of the material in the moulds and apply a light downward pressure to them. NB As the surfaces of the casts are tacky the washers will be retained in position.

12 Catalyse the mixture in the first container.

13 Pour enough of the catalysed mixture into the moulds just to cover the washers.

14 Examine the mixture in the moulds for air bubbles and remove any that may be present.

15 Allow the mixture to set.

16 Remove the casts from the moulds (p. 20 ‡ 29).

17 Finish the backs of the casts (p. 21 ‡ 34).

18 Pour opaque resin onto the backs of the two casts (p. 23 ‡ 37).

19 Pour 6 ml [6 cc] of resin into the fourth container and catalyse it.

20 Cover the contact surfaces of the cuff-link mountings with the catalysed resin and allow it to become tacky.

21 Position the cuff-link mountings, contact surfaces downwards, at the centres of the backs of the casts and apply a light downward pressure. NB Make sure the resin is tacky before the cuff-link mountings are positioned or they may slide over the surfaces of the casts and set off-centre.

22 Allow the cuff-link mountings to set in position.

Designing and Making Original Jewelry

57

The examples of jewelry illustrated in this book represent an attempt to discover a style that is true to the character of polyester resin.

We suggest that there are two ways in which you may design and make original pieces of jewelry. You may put the techniques described in this book to work in new ways (fig. 57) or you may develop new techniques and produce jewelry that is highly distinctive (figs. 58–62). Whichever course you choose, always remember that the art of using polyester resin is new and that there is much satisfaction to be gained by those of us who contribute to its development.

58

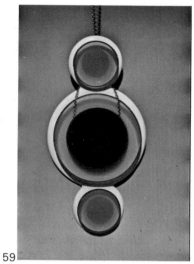

59

60

61

62

63

Suppliers

United Kingdom

Polyester Resin and Accessories

Isopon Inter Chemicals
Derbyshire House,
St Chad's Street,
London WC1H 8AH.

Strand Glass Co. Ltd.,
Brentwood Trading Estate,
Brentford,
Middlesex.

Trylon Ltd.,
Wollaston,
Northants,
NN9 7QJ.

Jeweller's Findings

W. G. Ball Ltd.,
Longton,
Stoke-on-Trent,
ST3 1JW.

Charles Cooper (Hatton Garden) Ltd.,
Wall House,
12 Hatton Wall,
Hatton Garden,
London EC1.

E. Gray & Son Ltd.,
12, 14 & 16 Clerkenwell Road,
London EC1

United States

Polyester Resin and Accessories

Polyproducts Corporation,
Order Dept., Room 25
13810 Nelson Avenue
Detroit, Michigan 48227.

Resin Coatings Corporation
14940 N.W. 25 Court,
Opa Locka, Florida 33054.

Jeweller's Findings

Allcraft Tool & Supply Co. Inc.
15 West 25th Street,
New York 36,
N.Y.

American Handicraft Co. Inc.,
20 West 14th Street,
New York.

The Evans Findings Co. Inc.,
55 John Street,
Providence,
Rhode Island.

T. B. Hagstoz & Son
709 Sansom Street,
Philadelphia, Pennsylvania, 19106.

Jeweller's findings may be bought in
most shops and departments of
large stores that deal in craft goods.